Exploring the Grand Canyon

Colleen Adams

The Rosen Publishing Group, Inc.
New York

Published in 2001 by The Rosen Publishing Group, Inc.
29 East 21st Street, New York, NY 10010

Copyright © 2001 by The Rosen Publishing Group, Inc.

All rights reserved. No part of this book may be reproduced in any form without permission in writing from the publisher, except by a reviewer.

Book Design: Haley Wilson

Photo Credits: Cover, pp. 1, 2–3, 22–23, 24 © Ron Thomas/FPG International; p. 4 © David Noble/FPG International; pp. 7, 16 © Corbis; p. 8 © Richard Price/FPG International; p. 11 © Gail Shumway/FPG International; p. 12 © Gary Randall/FPG International; p. 15 © Robert Stottlemyer/International Stock/Tom Till/International Stock/Scott T. Smith/Corbis; p. 19 © Bill Gillette/Stock, Boston/Picture Quest; p. 20 © Gerald French/FPG International.

ISBN: 0-8239-8163-0
6-pack ISBN: 0-8239-8565-2

Manufactured in the United States of America

Contents

A Natural Wonder	5
The Mighty Colorado	9
The Rocks Tell a Story	13
People of the Canyon	18
Grand Canyon National Park	21
Glossary	23
Index	24

A Natural Wonder

The Grand Canyon is one of the seven natural wonders of the world. It is about 277 miles long and up to 18 miles wide in some places. The canyon is one mile deep from the edge to the floor. It has steep cliffs, winding paths, and high peaks. <u>It</u> is made up of rocks that are many different shapes, sizes, and <u>colors</u>.

The Grand Canyon is in the northwest corner of Arizona. Most people visit the south side of the canyon, which is called the South Rim. The **climate** there is very hot and dry. It can get up to almost 120 degrees in the summer! The north side of the canyon is called the North Rim. The North Rim is about 1,200 feet higher than the South Rim. The climate there is cooler and wetter.

It can be as much as 25 degrees warmer at the bottom of the canyon than at the canyon's rim.

Native Americans have lived in and around the canyon for thousands of years. The first European **explorer** to find the canyon was Captain Garcia Lopez de Cardenas, who arrived there in 1540. Cardenas and his men were searching for gold. They spent several days trying to get through the canyon, but they ran out of supplies and had to give up. Few people knew about the Grand Canyon until people began to make maps of this part of the world in the mid-1700s.

In 1869, a **geologist** named John Wesley Powell traveled through the canyon with a team of men. They studied the rocks, plants, animals, and people who lived in the area of the canyon. The story of Powell's trip through the canyon soon spread, and more and more people wanted to visit the Grand Canyon.

The trip through the Grand Canyon was difficult for Powell (far left) and his men. It took them ninety-nine days!

The Mighty Colorado

You might wonder what could create such a big canyon. It is hard to believe, but the Colorado River did! About six million years ago, the Colorado River began to cut through the different layers of Earth's rock. Large amounts of water carried about 500,000 tons of mud, sand, and gravel downstream every day. This **sediment** in the river acted like sandpaper on the land. Over millions of years, the force of the water and the sediment chipped away at the rock and carved out the canyon.

Today, the Colorado River continues to wear away the canyon's layers of rock. Rain, wind, and changes in temperature also wear down the rock. Because of this, the Grand Canyon gets a little bigger every year.

The Colorado River still runs through the lowest part of the canyon. At 1,450 miles long, it is the longest river west of the Rocky Mountains.

The rocks in the Grand Canyon are many different colors. These colors seem to change from moment to moment as the sun moves across the sky. Most of the rocks are red, but some parts of the canyon are gray, green, and even pink. Some rocks are brown or purple. The rocks were not always these colors. Over many years, the Colorado River washed away minerals from different layers of rock. These minerals stained the rocks different colors.

Constant **erosion** has also helped create this rainbow of colors in the canyon. As water and wind wear away the rocks, different layers appear. These rocks change colors as more and more rock is worn away. The beautiful colors are one reason that millions of people visit the Grand Canyon each year.

The rocks in the Grand Canyon tell us that the area went through changes in weather and temperature over millions of years.

The Rocks Tell a Story

Scientists have learned a lot about Earth by studying the Grand Canyon. The canyon's layers show rocks from many different time periods. The oldest rocks are at the bottom of the canyon. They are mostly made of **schist** and **granite**. These rocks are about two billion years old. Above the granite are two kinds of rock known as **limestone** and **shale**. Some of these rocks are more than 500 million years old. These rocks formed in layers over time.

There are some places in the canyon where the layers of rock do not make sense. For example, one layer of **sandstone** is about 825 million years old, and right on top of it is a layer of sandstone that is only 545 million years old. Scientists believe that a sea covered the land during the 300 million years in between and washed away many layers of rock.

The canyon's different layers of rock show us what the conditions were like on Earth when the layers were being formed.

The different kinds of rock in the Grand Canyon give us many clues about what Earth used to be like. Limestone is a kind of rock that is formed underwater. If you look carefully at the limestone in the canyon, you might see seashells in the rock. The rock is made of bones, teeth, and shells of dead sea animals that once lived underwater. This tells us that a sea once covered the area.

Sandstone is another type of rock that is found in the canyon. It is made of sand. One layer of the canyon is made of wind-deposited sand, which tells us that the Grand Canyon had hot, dry weather conditions.

Another kind of rock called shale is formed by mud. When shallow water covered the land, it left mud that slowly hardened into shale.

Limestone, sandstone, and shale can all be found in the Grand Canyon's layers of rock.

The layers of rock in the Grand Canyon also tell us what kinds of plants and animals have lived in the area over time. Many rocks contain different kinds of **fossils**. A fossil is the hardened remains of a plant or animal, or the shape of a plant or animal that has hardened into stone.

There is a layer of limestone in the western part of the canyon that is much thicker than the layer of limestone in the eastern part of the canyon. Scientists have found a greater number of fossils in this thicker layer. Because of this, scientists believe that there was once more water in the western part of the canyon.

By studying fossils from the canyon, geologists have learned that some of the canyon was once covered by a deep sea.

People of the Canyon

Scientists have found many **artifacts** in the Grand Canyon. Some artifacts, such as dolls and carved figures, belonged to people who lived in the area more than 4,000 years ago. Scientists have also found the ancient **ruins** of people's homes.

A Native American group called the Ancestral Puebloans (once known as the Anasazi) lived near the canyon about 2,000 years ago. They lived in homes called cliff dwellings, which were built on or near the cliffs of the canyon. The Ancestral Puebloans probably left the area because the weather became too hot and dry. There was little water for growing food. Today, the Havasupai are the only Native Americans who still live in the canyon.

"Havasupai" means "people of the blue-green water." The Grand Canyon is a major part of the Havasupai's history.

Grand Canyon National Park

In 1919, the United States government made a large part of the Grand Canyon into a national park. This was done to **preserve** the canyon's beauty and wildlife. The government has paved roads around the rims of the canyon to make it easier for people to travel.

There are also trails leading down into the canyon. Some of these trails are very narrow. Although there are many routes across the canyon, there is only one continuous trail that crosses the canyon from the North Rim to the South Rim. It is called Kaibab Trail. The best way to travel on the canyon's trails is on foot or by mule. Mules carry visitors through the canyon. The mules are led by guides who tell visitors many interesting facts about how the canyon was formed.

Some trails are so narrow that there is barely enough room for one mule at a time!

The rocks of the Grand Canyon tell a wonderful story about how Earth changes over time. Many plants and animals now live in and around the canyon where once there were very few. There are coyotes, foxes, and deer, to name just a few. There are forests of pine, fir, and spruce on the North Rim. The hotter South Rim has plants that need less water, such as cactus and **yucca**.

The Colorado River continues to wash away pieces of rock from the bottom of the canyon. Harsh weather still wears down the rocks. This makes the canyon deeper and wider. We can only wonder what the canyon will look like in another billion years.

Glossary

artifact	An object created by a human being.
climate	The kind of weather a place has.
erosion	The process of being worn away a little at a time by water, air, or other forces.
explorer	A person who travels to new lands to find new things.
fossil	The hardened remains of a dead animal or plant that lived long ago.
geologist	A scientist who studies rocks.
granite	A very hard gray or pink rock that is often used in building.
limestone	A rock that is made mostly of the remains of shells and coral.
preserve	To keep something from harm or change.
ruins	Something that is very old and damaged.
sandstone	A kind of rock that is made mostly of sand.
schist	A kind of rock made of crystals.
sediment	The mud, sand, and gravel that are carried by wind or water.
shale	A rock formed from hardened clay or mud.
yucca	A plant found in dry, warm areas of North and Central America.

Index

A
animal(s), 6, 14, 17, 22
Arizona, 5
artifacts, 18

C
Cardenas, Garcia Lopez de, 6
cliff(s), 5, 18
climate, 5
Colorado River, 9, 10, 22
colors, 5, 10

E
Earth, 9, 13, 14, 22

F
fossils, 17

G
granite, 13

L
limestone, 13, 14, 17

M
mud, 9, 14

N
Native American(s), 6, 18
North Rim, 5, 21, 22

P
paths, 5
peaks, 5
plant(s), 6, 17, 22
Powell, John Wesley, 6

R
rock(s), 5, 6, 9, 10, 13, 14, 17, 22

S
sandstone, 13, 14
sea, 13, 14
sediment, 9
shale, 13, 14
South Rim, 5, 21, 22

W
water, 9, 10, 14, 17, 18, 22
wind, 9, 10